T0237599

Introduction to Engineering
A Starter's Guide With Hands-On
Analog Multimedia Explorations

© Springer Nature Switzerland AG 2022
Reprint of original edition © Morgan & Claypool 2008

All rights reserved. No part of this publication may be reproduced, stored in a retrieval system, or transmitted in any form or by any means—electronic, mechanical, photocopy, recording, or any other except for brief quotations in printed reviews, without the prior permission of the publisher.

Introduction to Engineering: A Starter's Guide With Hands-On Analog Multimedia Explorations
Lina J. Karam and Naji Mounsef

ISBN: 978-3-031-79317-2   paperback
ISBN: 978-3-031-79318-9   ebook

DOI: 10.1007/978-3-031-79318-9

A Publication in the Springer series

SYNTHESIS LECTURES ON ENGINEERING #6

Lecture #6

Series ISSN
ISSN: 1939-5221   print
ISSN: 1939-523X   electronic

Introduction to Engineering

A Starter's Guide With Hands-On Analog Multimedia Explorations

Lina J. Karam and Naji Mounsef
Arizona State University

SYNTHESIS LECTURES ON ENGINEERING #6

ABSTRACT

This lecture provides a hands-on glimpse of the field of electrical engineering. The introduced applications utilize the NI ELVIS hardware and software platform to explore concepts such as circuits, power, analog sensing, and introductory analog signal processing such as signal generation, analog filtering, and audio and music processing. These principals and technologies are introduced in a very practical way and are fundamental to many of the electronic devices we use today. Some examples include photodetection, analog signal (audio, light, temperature) level meter, and analog music equalizer.

KEYWORDS

Analog, circuits, signal processing, filtering, audio, music, analog sensing, NI ELVIS, hands-on applications.

Foreword

This volume provides a hands-on glimpse of the field of electrical engineering. The introduced applications utilize the NI ELVIS platform to explore concepts such as: circuits, power, analog sensing, and introductory analog signal processing such as signal generation, analog filtering, audio and music processing. These principals and technologies are introduced in a very practical way and are fundamental to many of the electronic devices we use today. Some examples include photodetection, analog signal (audio, light, temperature) level meter, and analog music equalizer.

The provided hands-on applications can be easily expanded into longer-duration design projects. Examples of such design projects can be found at http://www.fulton.asu.edu/~karam/introeng.

National Instruments Part Number Information

778748-02	NI ELVIS/PCI-6251 DAQBoard Bundle (Academic Use Only)
780378-01	NI ELVIS II Basic Bundle (Academic Use Only)

Contents

CHAPTER 1

Getting Familiar with
NI ELVIS

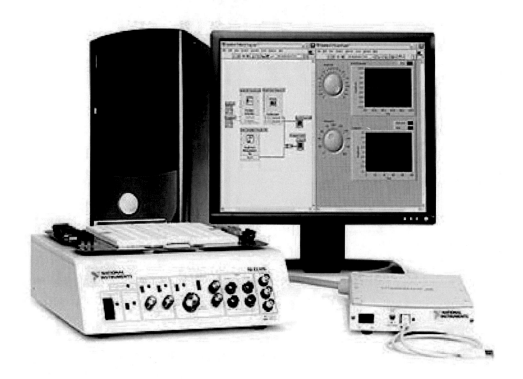

1.1 OVERVIEW

This experiment is an introduction to the NI ELVIS platform and its many features that exist in both hardware and software. It begins with an introduction to the NI ELVIS workspace and eventually steps through the important instruments available in the NI ELVIS software. Protoboards will be used with the common analog components to test the different features of NI ELVIS.

1.2   BACKGROUND
1.2.1   NI ELVIS

The Educational Laboratory Virtual Instrumentation Suite (ELVIS) produced by National Instruments (NI) is designed to simulate several pieces of electronic laboratory equipment. It eliminates the need for bulky equipment in the lab. It also allows for the design of customized instrumentation that can be used for specific projects.

NI ELVIS was designed to function as a three-part system. The NI ELVIS workstation interfaces with National Instruments LabVIEW software and a National Instruments data acquisition device to perform measurements and transmit signals. Both LabVIEW and a National Instruments data acquisition device are required for the NI ELVIS to operate due to the workstation's dependency on these components to make measurements and send control signals. The integration of these three parts makes NI ELVIS a powerful, highly customizable measurement and experimentation platform. The entire NI ELVIS system is built on top of NI's Labview software and hardware, and is comprised of three main components as shown and described below:

1. ELVIS software on the PC, which provides a graphical interface for several electronic instruments. These instruments are discussed in detail throughout the rest of this guide.
2. The PC connects to the bench-top workstation through the data acquisition board. This board has both analog and digital input and output lines.
3. The bench-top workstation provides electrical connections for the user to interface with circuits. These provide inputs and outputs for the virtual instrumentation provided by the ELVIS software. There is a prototyping board (breadboard) on the top of the workstation with several holes connected through the data acquisition board. There is also a front interface on the workstation with controls for a few select instruments. The workstation and its connections will be described in detail in the next sections.

1.2.2   NI ELVIS Workstation

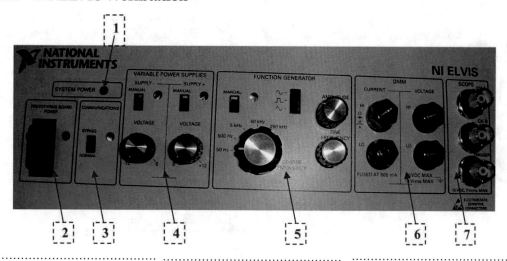

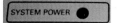

1-   System Power LED

Indicates whether the NI ELVIS is powered on.

2-   Prototyping Board Power

Controls the power to the prototyping board.

3-   Communications

Requests disabling software control of the NI ELVIS.

4-   Variable Power Supplies Control

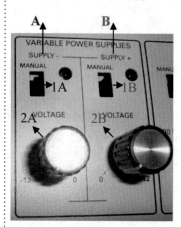

A) Supply- Controls

    1A: **Manual Switch**—Controls whether the negative supply is in Manual mode or Software mode.

    2A: **Voltage Adjust Knob**—Controls the output of the negative supply. The negative supply can output between –12 and 0 V.

B) Supply+ Controls

    1B: **Manual Switch**—Controls whether the positive supply is in Manual mode or Software mode.

    2B: **Voltage Adjust Knob**—Controls the output of the positive supply. The positive supply can output between 0 and +12 V.

5-   Function Generator

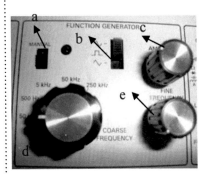

a: **Manual Switch**—Controls whether the function generator is in Manual mode or Software mode.

b: **Function Selector**—Selects what type of waveform is generated. NI ELVIS can generate sine, square, or triangle waves.

c: **Amplitude Knob**—Adjusts the peak amplitude of the generated waveform.

d: **Coarse Frequency Knob**—Sets the range of frequencies the function generator can generate.

e: **Fine Frequency Knob**—Adjusts the output frequency of the function generator.

6-   DMM Connectors

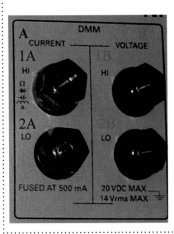

A) CURRENT Banana Jacks

1A: **HI**—The positive input to all the Digital Multi Meter (DMM) functionality, except measuring voltage.

1B: **LO**—The negative input to all the Digital Multi Meter (DMM) functionality, except measuring voltage.

B) VOLTAGE Banana Jacks

1B: **HI**—The positive input for voltage measurements.

2B: **LO**—The negative input for voltage measurements.

7-   Oscilloscope Connectors

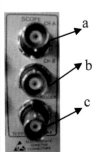

a: **CH A BNC Connector**—The input for channel A of the oscilloscope.

b: **CH B BNC Connector**—The input for channel B of the oscilloscope.

c: **Trigger BNC Connector**—The input to the trigger of the oscilloscope.

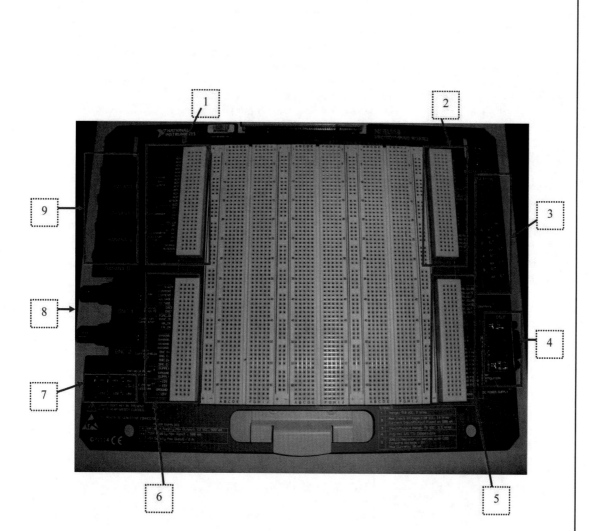

1   Analog Input Signals/Oscilloscope/Programmable Function I/O Signal Rows

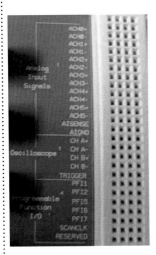

ACH<0..5>+ General AI Analog Input Channels 0 through 5 (+)—Positive differential input to the AI channels.

ACH<0..5>– General AI Analog Input Channels 0 through 5 (–)—Negative differential input to the AI channels.

AISENSE General AI Analog Input Sense—Reference for the analog channels in nonreferenced single-ended (NRSE) mode.

AIGND General AI Analog Input Ground—AI ground reference for the DAQ device. This ground signal is not tied to the NI ELVIS GROUND signals.

CH <A..B>+ Oscilloscope Oscilloscope Channels A and B (+)—Positive input for the oscilloscope channels.

CH <A..B>– Oscilloscope Oscilloscope Channels A and B (–)—Negative input for the oscilloscope channels.

TRIGGER Oscilloscope Oscilloscope Trigger—Trigger input for the oscilloscope, referenced to AIGND.

PFI <1..2>, PFI <5..7> Programmable Function I/O Programmable Function Input (PFI) 1 through 2 and 5 through 7—Programmable function I/O of the DAQ device.

SCANCLK Programmable Function I/O Scan Clock—Connected to the SCANCLK pin of the DAQ device.

RESERVED Programmable Function I/O Connected to the EXTSTROBE* pin of the DAQ device.

2   Digital Input/Output Signal Rows

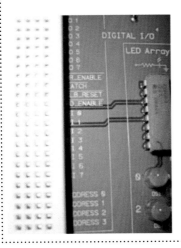

DO <0..7> DIO Digital Output Lines 0 through 7—Output of the write bus.

WR ENABLE DIO Write Enable—Output indicating data is being written to the write bus.

LATCH DIO Latch—Output that pulses after data is ready on the write bus.

GLB RESET DIO Global Reset—Output indicating global digital reset.

RD ENABLE DIO Read Enable—Output indicating data is being read from read bus.

DI <0..7> DIO Digital Input Lines 0 through 7—Output of read bus.

ADDRESS <0..3> DIO Address Lines 0 through 3—Output of address bus.

3 LED Array

4 DSUB

7 Power LEDS

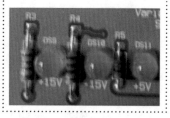

9 Banana Jack Connectors

8 BNC Connectors

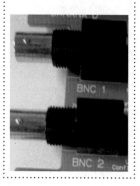

5 Counter/Timer, User-Configurable I/O, and DC Power Supply Signal Rows

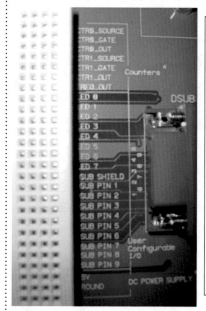

CTR0_SOURCE Counters Counter 0 Source
Connected to the GPCTR0_SOURCE pin on the DAQ device.
CTR0_GATE Counters Counter 0 Gate
Connected to the GPCTR0_GATE pin on the DAQ device.
CTR0_OUT Counters Counter 0 Output
Connected to the GPCTR0_OUT pin on the DAQ device.
CTR1_GATE Counters Counter 1 Gate
Connected to the GPCTR1_GATE pin on the DAQ device.
CTR1_OUT Counters Counter 1 Output
Connected to the GPCTR1_OUT pin on the DAQ device.
FREQ_OUT Counters Frequency Output
Connected to the FREQ_OUT pin on the DAQ device.
LED <0..7>
User Configurable I/O LEDs 0 through 7—Input to the LEDs.
DSUB SHIELD User Configurable I/O D-Sub Shield
Connection to D-Sub shield.
DSUB PIN <1..9>
User Configurable I/O D-Sub Pins 1 through 9
Connection to D-Sub pins.

6- DMM, AO, Function Generator, User-Configurable I/O, Variable Power Supplies

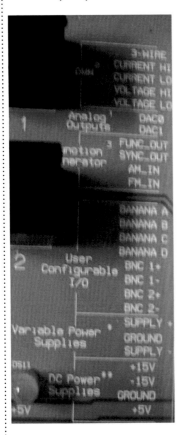

3-WIRE DMM Three Wire
Voltage source for the DMM for three-wire transistor measurements.
CURRENT HI DMM Positive Current
Positive input for the DMM for all measurements besides voltage.
CURRENT LO DMM Negative Current
Negative input for the DMM for all measurements besides voltage.
VOLTAGE HI DMM Positive Voltage
Positive input for the DMM voltmeter.
VOLTAGE LO DMM Negative Voltage
Negative input for the DMM voltmeter.
DAC<0..1>
General AO Analog Channel Output for Channels 0 through 1—Outputs of the DAQ device D/A converters (DACs).
FUNC_OUT Function Generator Function Output
Output of the Function Generator.
SYNC_OUT Function Generator Synchronization Output
TTL signal of the same frequency as the output as the FUNC OUT pin.
AM_IN Function Generator Amplitude Modulation Input
Input to the amplitude modulator for the Function Generator.
FM_IN Function Generator Frequency Modulation Input
Input to the frequency modulator for the Function Generator.
BANANA <A..D> User Configurable I/O Banana Jacks A through D
Connects to the banana jacks pins.
BNC <1..2>+ User Configurable I/O BNC Connectors 1 and 2 (+)
Connects to the BNC pins.
BNC <1..2>− User Configurable I/O BNC Connectors 1 and 2 (−)
Connects to the BNC pins.
SUPPLY+ Variable Power Supplies Positive
Output of 0 to 12 V variable power supply.
SUPPLY− Variable Power Supplies Negative
Output of −12 to 0 V variable power supply.
Ground
Prototyping board ground. These two ground signals are tied together.
+15 V DC Power Supplies +15 V Source
Output of fixed +15 V power supply, referenced to the NI ELVIS GROUND signal.
−15 V DC Power Supplies −15 V Source
Output of fixed −15 V power supply, referenced to the NI ELVIS GROUND signal.
+5V DC Power Supplies +5V Source
Output of fixed +5 V power supply, referenced to the NI ELVIS GROUND signal.

1.2.3 NI ELVIS Protoboard

A protoboard is used to make up *temporary circuits* for testing or to try out an idea. No soldering is required so it is easy to change connections and replace components. Parts will not be damaged so they will be available for reuse afterwards. Breadboards have many tiny sockets (called "holes") arranged on a 0.1-in. grid. The leads of most components can be pushed straight into the holes. ICs are inserted across the central gap with their notch or dot to the left. The connections on the red and black +/− sections are made in vertical columns. The connections in the other sections are made in sets of five horizontal holes separated by the valley in the middle.

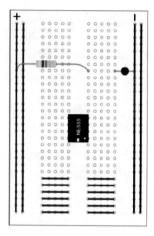

On the ELVIS protoboard, there are two types of holes:

Breadboard holes: The bulk of the breadboard is made up of holes that are not connected to the data acquisition board and the PC. These connections are just like a normal breadboard that can be found in any circuits laboratory.

Special holes on the breadboard: These are the holes that are found on the far left and far right of the breadboard. They consist of several sets of four horizontal holes that are also connected to the PC for special functions.

The figure below shows the layout of the interconnections inside the board.

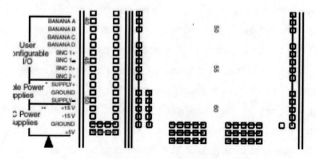

Figure: Layout of the Prototyping Board

1.3   BASIC CIRCUIT COMPONENTS

There are three basic, analog circuit components: the resistor (R), capacitor (C) and inductor (L). In this chapter, we will be using only resistors and capacitors.

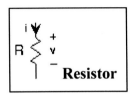

The resistor is far and away the simplest circuit element. In a resistor, the voltage is proportional to the current, with the constant of proportionality R known as the resistance. Resistance has units of ohms, denoted by Ω, named for the German electrical scientist Georg Ohm. As the resistance approaches infinity, we have what is known as an open circuit: No current flows but a nonzero voltage can appear across the open circuit.

As the resistance becomes zero, the voltage goes to zero for a nonzero current flow. This situation corresponds to a short circuit. A superconductor physically realizes a short circuit.

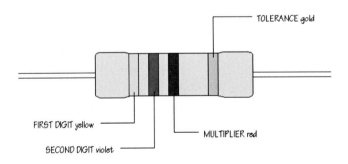

The first band of a resistor is interpreted as the *first digit* of the resistor value. For the resistor shown below, the first band is yellow, so the first digit is 4. The second band gives the *second digit*. This is a violet band, making the second digit 7. The third band is called the *multiplier* and is not interpreted in quite the same way. The multiplier tells you how many zeros you should write after the digits you already have. A red band tells you to add two zeros. The value of this resistor is therefore *4 7 0 0* ohms, that is, 4,700 Ω, or 4.7 kΩ. Work through this example again to confirm that you understand how to apply the color code given by the first three bands. The remaining band is called the *tolerance* band. This indicates the percentage accuracy of the resistor value. Most carbon film resistors have a gold-colored tolerance band, indicating that the actual resistance value is with + or − 5% of the nominal value.

Each color represents a number according to the following scheme:

Number	Colour
0	black
1	brown
2	red
3	orange
4	yellow
5	green
6	blue
7	violet
8	grey
9	white

Tolerance	Colour
±1%	brown
±2%	red
±5%	gold
±10%	silver

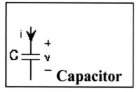

Capacitor

The capacitor stores charge; as current is the rate of change of charge, integrating the capacitor's v–i relation yields $q = Cv$. The charge stored in a capacitor is proportional to the voltage. The constant of proportionality, the capacitance, has units of farads (F), and is named for the English experimental physicist Michael Faraday. If the voltage across a capacitor is constant, then the current flowing into it equals zero.

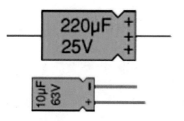

Large capacitors will usually have their value printed on them. For instance, you may see 10 µF printed on the capacitor. In this case, you have a 10-microfarad capacitor. Also watch out for the decimal point. The capacitor may have 0.47 µF printed on it not 47 µF.

 To save space, some capacitors use a numeric code to represent the actual capacitor value. This capacitor code works similarly to the resistor color code. You may have a capacitor that has 102 printed on it. Often, it will also have 100 V printed below it. The 102 represents the capacitor value, and the 100 V tells us it is rated for 100 volts.

To read the capacitor value, take the first two digits as the first and second significant digits. The third digit is a multiplier. This multiplier is the number of zeros to add after the first two digits. The resulting value you get is in picofarads which you can then convert to microfarads if necessary.

For example, the 102 capacitor decodes to 1,000 pF, or a 1-nF capacitor (1,000 pF = 1 nF). Components:

- Resistors: 1k, 2.2k
- Capacitors: 1uF

1.4   EXPERIMENT

Now that you have an idea about NI ELVIS, you are ready to begin the experiment.

1.4.1   Measuring Component Values

Complete the following steps to measure component values:

- Connect two banana-type leads to the DMM current inputs on the workstation front panel.
- Connect the other ends to one of the resistors.
- Launch NI ELVIS by double-clicking the NI ELVIS desktop icon. After initializing, the suite of LabVIEW software instruments opens as shown in Figure 1.2.

Select *Digital Multimeter*. A message box will open. Read the message and click **OK**. Click the **Null** button.

You can use the Digital Multimeter Software Front Panel (SFP) for a variety of operations. The notation DMM[X] is used to signify the X operation. Click the Ω (ohm) button to use the Digital Ohmmeter function DMM[Ω] to measure different resistor values R_1, R_2.

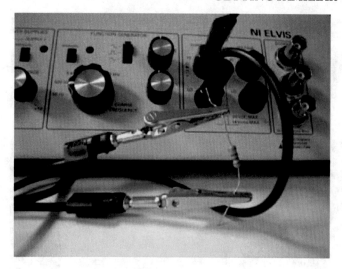

FIGURE 1.1: Measuring resistor value using the ELVIS DMM.

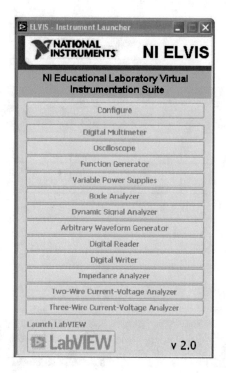

FIGURE 1.2: NI ELVIS instrument launcher.

FIGURE 1.3: NI ELVIS digital multimeter SFP.

1.   Fill in the following data:

RESISTOR	COLOR CODE	THEORETICAL VALUE	MEASURED VALUE
R_1			
R_2			

2.   Click the ┤├ (capacitor) button and measure a capacitor C with DMM[C] using the same leads. Fill in the following data:

CAPACITOR	CODE	THEORETICAL VALUE	MEASURED VALUE
C			

1.4.2   Building a Voltage Divider Circuit on the NI ELVIS Protoboard

Complete the following steps to build a voltage divider:

3.   Using the two resistors, R_1 and R_2, assemble the following circuit on the NI ELVIS protoboard.

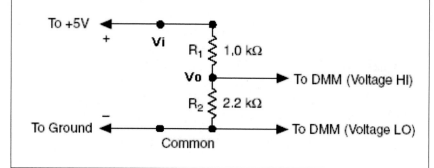

- Connect the input voltage, V_i, to the [+5V] pin socket.
- Connect the common to the NI ELVIS [Ground] pin socket.
- Connect the external leads to the DMM voltage inputs (HI) and (LO) on the front panel of the NI ELVIS workstation. Notice that there are separate input leads for voltage and impedance/current measurements.

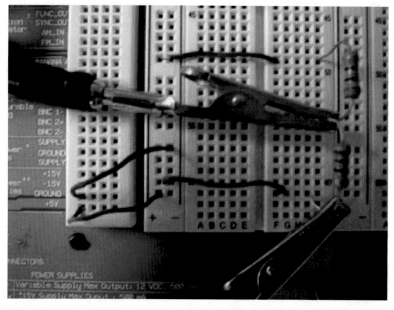

FIGURE 1.4: Measuring voltage for the voltage divider Circuit.

- Check the circuit, and then apply power to the protoboard by switching the Prototyping Board Power switch to the up position. The three power indicator LEDs, +15V, −15V, and +5V, should now be lit.
- Move the front panel leads back to the DMM(VOLTAGE) inputs and select DMM[V].

4.   Connect the DMM front panel leads to, Vi, and measure the input voltage using the DMM[V].

According to circuit theory, the output voltage V_o is as follows:

$$V_o = V_i * R_2 / (R_1 + R_2)$$

5.   Using the previously measured values for R_1, R_2, and V_i, calculate V_o.
Next, use the DMM[V] to measure the actual voltage, V_o.
V_o(calculated) _____ V_o(measured) _____
Compare the two values.

1.4.3   Using the DMM to Measure Current
According to Ohms law, the current (I) flowing in the previous circuit is equal to V_o / R_2.

6.   Using the measured values of V_o and R_2, calculate the current.

7.   Measure the current by moving the external leads to the workstation front panel DMM (Current) inputs HI and LO. Connect the other ends to the circuit, as shown in the following figure.

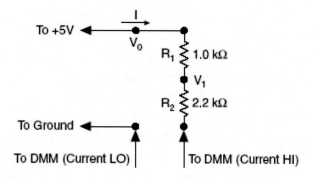

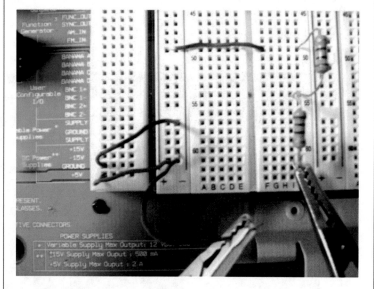

FIGURE 1.5: Measuring current.

8.   Select the function DMM[A] and measure the current.

I (calculated) _____

I (measured) _____

Compare the two values.

1.4.4   Using the Power Supply

Complete the following steps to take a measurement using the variable power supply:

- From the NI ELVIS Instrument Launcher, select *Variable Power Supplies*. There are two controllable power supplies with NI ELVIS: 0 to −12 V and 0 to +12 V, each with a 500-mA current limit.
- On the NI ELVIS workstation, slide the VPS+ switch to *Manual*.
- Connect the leads from [VPS+] and [Ground] sockets of the NI ELVIS prototyping board to the workstation DMM voltage inputs.
- Select DMM[V].
- Rotate the manual VPS knob on the workstation and observe the voltage change on the DMM[V].
- Slide the workstation switch for VPS+ down (not Manual). Now you can use the virtual VPS controls on the computer screen.
- Click and drag the virtual knob to change the output voltage.

FIGURE 1.6: Power supply SFP.

1.4.5   Using the Function Generator and Oscilloscope

Complete the following steps to build and test an RC circuit:

- On the workstation protoboard, build a voltage divider circuit using a 1-uF capacitor and a 1-k resistor.

9. Connect the RC circuit inputs to [FUNCOUT] and [Ground] pin sockets, as shown in the following figures.

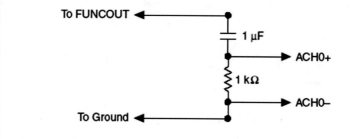

The power supply for an AC circuit is often a function generator, which you can use to test the RC circuit.

FIGURE 1.7: RC circuit.

From the NI ELVIS Instrument Launcher, select *Function Generator*.
The FGEN SFP has the following controls, which you can use to do the following:

- set the Frequency by decades (Course) and by Hz (Fine)
- select the waveform type (Sine, Square, or Triangle)
- select the waveform amplitude

All of these controls are also available on the workstation front panel. You can select them by sliding the workstation front panel function generator switch to Manual. As with the Variable Power Supply, manual control turns on the green LED display on the SFP and grays out the virtual controls.

You can use the oscilloscope to analyze the voltage signals of the RC circuit by completing the following steps:

- From the NI ELVIS Instrument Launcher, select *Oscilloscope*. The oscilloscope SFP is similar to most oscilloscopes, but the NI ELVIS oscilloscope can automatically connect inputs to a variety of sources.

There are measurements options, such as frequency and Amplitude P-P, which are accessed by clicking the *MEAS* buttons for either channel A or B. Measurements show up at the bottom of the oscilloscope screen. You can activate the cursors for channel A or B to make amplitude and time measurements.

Connect the workstation BNC SCOPE input CH B to the 1-k resistor.
Set function generator to the following state:

1. Turn ON
2. Sine Waveform
3. Peak amplitude = 1 V
4. Frequency = 10 Hz

Set Oscilloscope to the following state:

5. Turn ON both displays
6. Channel A to FGEN FUNC_OUT.
7. Channel B to BNC/BOARD CH B.
Time base = 100 ms

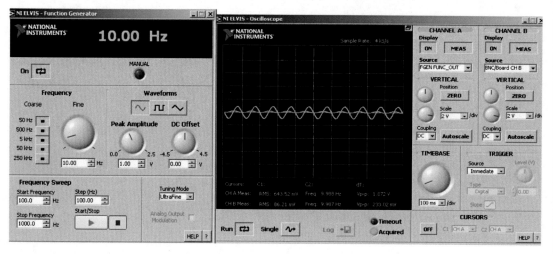

FIGURE 1.8: Function generator and oscilloscope SFPs.

The ratio of the amplitude on channel B to the amplitude on channel A defines the circuit gain at a particular frequency. Because there is no amplifier in the circuit, the gain is less than one. By looking at different frequencies, you can get a feel for the frequency characteristics of the RC passive filter circuit.

10. Read the following values of the oscilloscope

V p-p for frequency 1 = 10 Hz of Channel A _____

V p-p for frequency 1 = 10 Hz of Channel B _____

V p-p for frequency 1 = 10 kHz of Channel A _____

V p-p for frequency 1 = 10 kHz of Channel B _____

Try some frequencies between 10 Hz and 10 kHz and change time base on oscilloscope for proper viewing

What is the function of this circuit?

1.4.6   Photodetector Application

A photodetector is a circuit that changes resistance depending on the amount of light present. In this section, we will build the photodetection circuit with an infrared (IR) emitter and detector. The description of the emitter and detector is as follows:

Emitter
Wavelength at peak emission = 0.950 μm

Detector
Peak sensitivity wavelength = 0.850 μm
Spectral bandwidth range = 0.620–0.980 μm
Very high resistance when IR signal is absent
Low resistance when IR signal is present

Figure 1.9 is a representation of the electromagnetic spectrum with respect to wavelength. The emitter and detector fall into the lower end of the infrared spectrum, which is just above visible light.

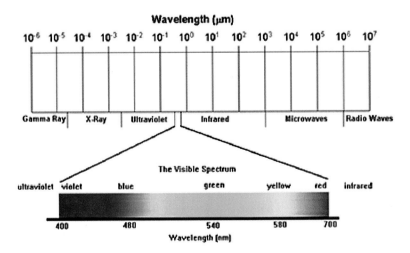

FIGURE 1.9: Infrared wavelength.

Complete the circuit as follows:

11. Comment on the similarities of the voltage divider circuit and the photodetector circuit. What are some applications of this circuit?

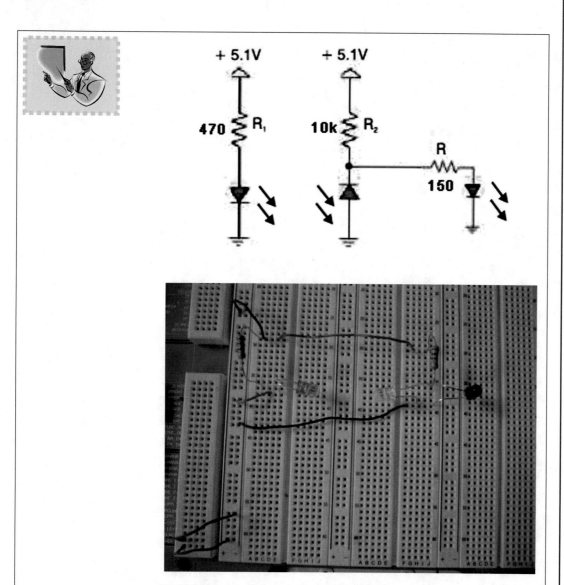

FIGURE 1.10: Infrared detector circuit.

CHAPTER 2

Analog Signal Level Meter Using LEDs

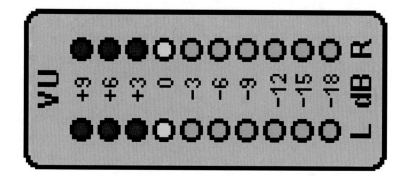

2.1 OVERVIEW

In this chapter, you will be using analog components such as LEDs and comparators to display the level of an analog signal, such as an audio, temperature or light signal.

2.2 BACKGROUND
2.2.1 Diodes

A *diode* is a component that allows an electric current to flow in one direction, but blocks it in the opposite direction.

It needs 0.7 V difference between its positive and negative terminals to operate. The negative side (cathode) of a diode is indicated by a gray dash.

Cathode

Anode

The symbol of a diode is shown below:

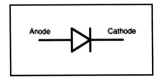

The first use for the diode was the demodulation of amplitude modulated (AM) radio broadcasts. However, they are most widely used in rectifiers, where they are used to convert alternating current (AC) electricity into direct current (DC).

2.2.2 Light-Emitting Diodes

Light-emitting diodes (LEDs)—small colored lights available in any electronics store—are ubiquitous in modern society. They are the indicator lights on our stereos, *automobile* dashboards, and microwave ovens. Numeric displays on clock radios, digital watches, and calculators are composed of bars of LEDs. LEDs also find applications in telecommunications for short range optical signal transmission such as TV remote controls. They have even found their way into jewelry and clothing—witness sun visors with a series of blinking colored lights adorning the brim. The inventors of the LED had no idea of the revolutionary item they were creating. They were trying to make lasers, but on the way, they discovered a substitute for the *light bulb*.

The *negative* side of an LED lead is indicated by the *shorter* of the two wires extending from the LED. The negative lead should be connected to the negative terminal of a battery. LEDs operate at relative low voltages between about 1 and 4 V, and draw currents between about 10 and 40 mA. Voltages and currents substantially above these values can destroy a LED chip.

The symbol of a LED is shown below:

2.2.3 Comparators

A comparator is the simplest circuit that moves signals between the analog and digital worlds. What does a comparator do? Simply put, a comparator compares two analog signals and produces a one-bit digital signal. The symbol for a comparator is shown below.

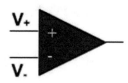

Way of operation:

If $V+ >= V-$ the comparator output voltage will be 5 V
If $V+ < V-$ the comparator output voltage will be 0 V

The LM339 comparator IC consists of four independent voltage comparators designed to operate from a single power supply over a wide voltage range.

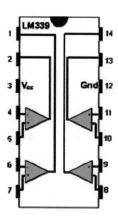
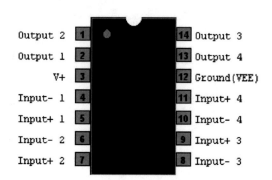

Components:

Resistor: 100k, 22k, 3k, 2.2k, 1.6k, Ten 1k, 820, 620, 390
Capacitor: 1u
Diode: 1N4002
Microphone/thermosensor/photosensor

2.3   EXPERIMENT

Now that you have an idea about LED's operations, you are ready to begin the experiment and implement an audio level meter using LEDs.

2.3.1   Application 1: Controlling 1 LED With One Comparator Using DC Power Supply

The first step to do is to test the operation of the comparator.

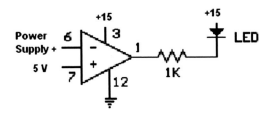

1. Construct the circuit and vary the Power supply + and notice the values of the Power Supply + for which the LED turns on and off. Comment.

2.3.2   Application 2: Converting Varying Signal (AC) to Constant Signal (DC)

In real life, the signal coming into pin 6 may not be constant, like the power supply signal, which is DC (constant). Therefore, the signal will be constantly fluctuating about the reference voltage to be compared with, and the LED at the output will be turning on and off all the time, which is not the way it should be operating. Hence, one should convert the varying signal to a constant DC signal. To do that, you use a rectifier circuit which is shown below.

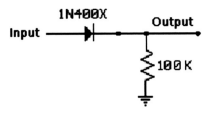

Keep the circuit that you constructed in application 1 and on the side, construct the circuit shown above and connect the ground to the corresponding socks on the ELVIS workstation and connect a sinewave using the function generator (Func_Out) of ELVIS to the Input.

2. Using the oscilloscope, observe the input and output at the same time. Describe the output. Vary the amplitude and the frequency of the sinewave and notice the output. Describe the output. Is the output constant or varying? Comment.

Add a capacitor in parallel with the resistor as shown below:

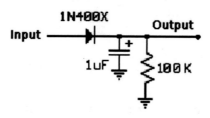

3. Observe on the oscilloscope the input and output at the same time. Vary the amplitude and frequency of the sinewave and notice the output. Describe the output. Is the output constant? Comment.

2.3.3   Application 3: First Stage of a Signal Level Meter

Replace the power supply in Application 1 by the circuit you built in Application 2. Replace the +5 V connected to pin 7 by a voltage divider circuit composed of 22k and 3k resistors connected to +12 V that you obtain from the Power Supply +. The resulting circuit should look as shown below.

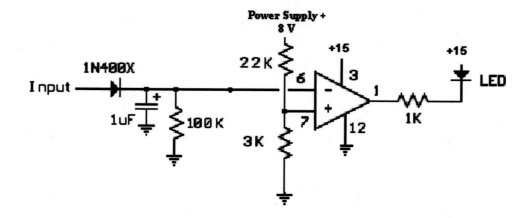

This circuit is the first stage of constructing an eight-LED audio level meter. This stage consists of one LED and one comparator. The circuit uses a comparator to illuminate an LED indicating high volume level.

4. Construct the circuit and connect the ground and the +15V DC to the corresponding socks on the ELVIS workstation. Observe the circuit in detail and explain its operation.

The comparator is biased at a specific voltage set by the voltage divider so that the LED turns on at a specific level.

5. To obtain the voltage at which the comparator is biased, measure the voltage at pin 7 and compare it with the theoretical value. Use the formula of a voltage divider circuit from Chapter 3.

6. Connect a function generator using ELVIS to the Input and vary the amplitude to see how it affects the LED status. What is the value at pin 6 at which the LED turns on? What is the corresponding amplitude of the input? Comment.

2.3.4   Application 4: Eight-LED Signal Level Meter

The eight-LED signal level meter is similar to the circuit you just constructed, but instead of comparing the incoming voltage with one voltage level, you will be comparing it with different voltage levels.

To create different voltage levels, construct the voltage divider circuit shown below:

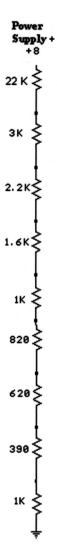

Power
Supply +
+8

22 K

3K

2.2K

1.6K

1K

820

620

390

1K

7. Measure the voltages at each node of the resistors. Compare with the theoretical values.

The next step is to connect a resistor and a LED to each of the outputs of the comparators:

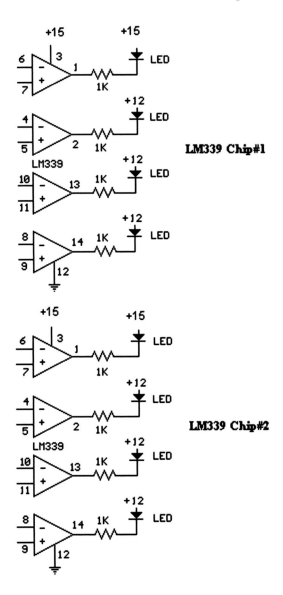

The LEDs should turn on if the incoming signal at the negative input of each comparator (e.g., 6) is higher than the voltage level at the corresponding positive input of each comparator (e.g., 7). For varying AC signals, such as audio signals, the incoming signal that will be used will be the output of the rectifier circuit that you constructed in Application 2. Connect it to each of the negative inputs of the comparators as shown below:

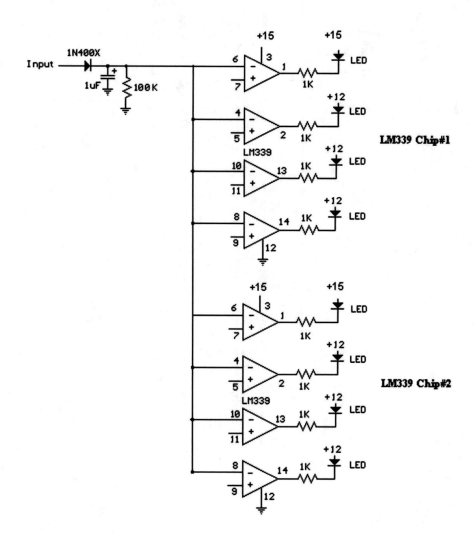

The last step is to connect different voltage levels at each of the positive inputs in decreasing order. The highest voltage should go to pin 7 of the first comparator and the lowest voltage will go to the pin 9 of the last comparator. The different voltage levels will be obtained from the voltage divider circuit you built previously as follows:

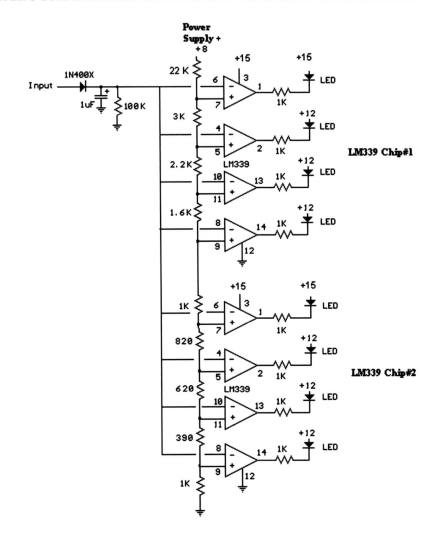

8. Connect a function generator using ELVIS to the Input and vary the amplitude to see how it affects the LEDs. Explain the operation of the circuit. Measure the voltage at the output of the rectifier circuit for which each LED turns on.

9. Test the circuit by using as input a microphone and notice how the LEDs change depending on the level of your voice.

Note: You can replace the microphone, the diode and the capacitor, with a photocell or a thermo-sensor to measure the light or temperature.

·   ·   ·   ·

CHAPTER 3

Noise Removal Using Analog Filters

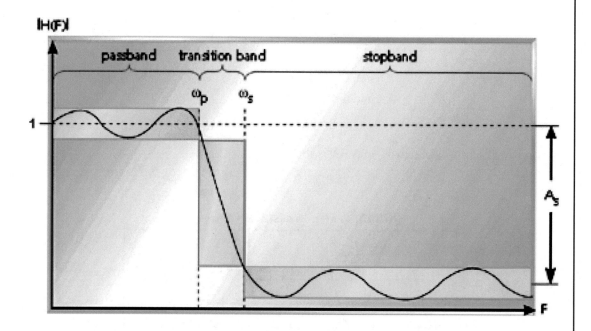

3.1   OVERVIEW

In this chapter, you will be using a few capacitors and resistors along with op amps to implement analog filters and get an idea about the difference with digital filters. NI ELVIS will be used to measure the characteristics of lowpass, highpass, and bandpass filters.

3.2   BACKGROUND

The most common representation of signals and waveforms is in the time domain. However, many signal analysis techniques work only in the frequency domain. When it is first introduced, the concept of the frequency domain representation of a signal is somehow difficult to understand. This section attempts to explain the frequency domain representation of signals. The frequency domain is simply another way of representing a signal.

3.3 SINUSOIDAL SIGNALS AND FREQUENCY

A *sinusoid* is any function of time having the following form:

$$x(t) = A \sin(2\pi f t + \phi)$$

where

$$A = \text{peak amplitude (nonnegative)}$$
$$t = \text{time (s)}$$
$$f = \text{frequency (Hz)}$$
$$\phi = \text{initial phase(radians)}$$

An example of a sinusoidal signal (sinusoid) is shown below:

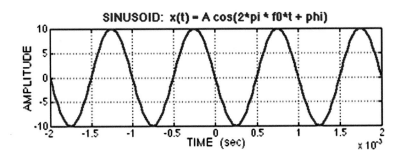

The term "peak amplitude" is often shortened to "amplitude."

The "phase" of a sinusoid normally means the "initial phase." Another term for initial phase is *phase offset.*

Note that Hz is an abbreviation for *hertz* which physically means *cycles per second.*

Although we only examined a sinusoidal waveform, it is relevant to all waveforms because any non-sinusoidal waveforms can be expressed as the sum of various sinusoidal components.

The following two sinusoids have frequencies of 300 and 500 Hz, respectively:

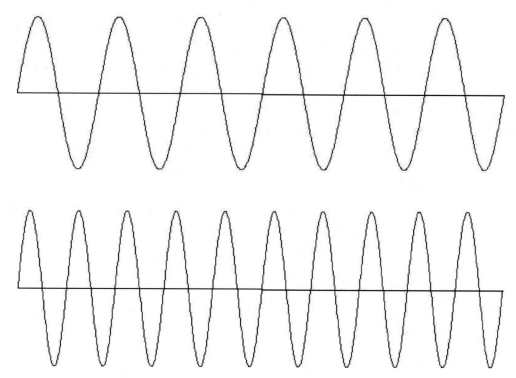

They can be added together to produce a more complex waveform:

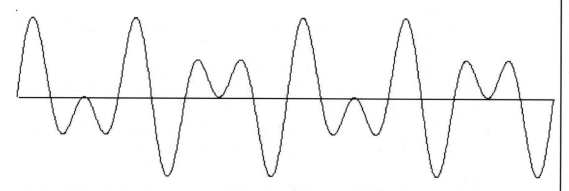

This is very important because, in general, *one can represent any complex continuous wave as a combination of simple sine waves with different frequencies*. In other words, any signal (voice, music . . .) is composed of different frequencies. For example, the voice signal is composed of frequencies ranging from 300 to 3,300 Hz, while music frequencies range from 20 Hz to 20 kHz.

One can take advantage of this fact to eliminate unwanted frequencies, as it will be shown in this experiment.

The frequency-domain display shows how much of the signal's energy is present as a function of frequency; in other words, for each frequency value, the frequency-domain representation displays the amplitude (amplitude spectrum) and phase (phase spectrum) of the sinusoidal component present in the signal at that frequency. For a simple signal such as a sine wave, the frequency domain representation does not usually show us much additional information. However, with more complex signals, the frequency domain gives a more useful view of the signal.

To summarize, any periodic waveform can be decomposed into a set of sinusoids with different amplitudes, frequencies, and phases. The process of doing this is called Fourier analysis, and the result is a set of amplitudes, phases, and frequencies for each of the sinusoids that make up the complex waveform. Adding these sinusoids together again will reproduce exactly the original waveform. A plot of the amplitude of a sinusoid against frequency is called an amplitude spectrum, and a plot of the phase against frequency is called phase spectrum.

3.4   WHAT IS A FILTER?

A filter is a device that accepts an input signal, and passes or amplifies selected frequencies while it blocks or attenuates unwanted ones.

For example, a typical phone line acts as a filter that limits frequencies to a range considerably smaller than the range of frequencies human beings can hear. That is why listening to CD-quality music over the phone is not as pleasing to the ear as listening to it directly using a CD player.

Filters can be *analog* or *digital*.

A *digital* filter takes a digital input, gives a digital output, and consists of digital components. In a typical digital filtering application, software running on a digital signal processor (DSP) reads input samples from an analog-to-digital (A/D) converter, performs discrete mathematical manipulations for the required filter type to possibly eliminate some frequencies, and outputs the result via a digital-to-analog (D/A) converter.

An *analog* filter, which will be covered here in this chapter, by contrast, operates directly on the analog inputs and is built entirely with analog components, such as resistors, capacitors, and inductors.

The *frequency response* of a filter is the measure of the filter's response (filter output) to a sinusoidal signal of varying frequency and unit amplitude at its input.

There are many filter types, but the most common ones are *lowpass, highpass, bandpass,* and *bandstop.* They are shown below along with their frequency response.

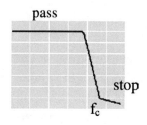

A *lowpass* filter allows only low frequency signals (below some specified cutoff frequency, f_c) through to its output, so it can be used to eliminate high frequencies.

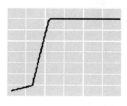

A *highpass* filter does just the opposite, by rejecting only frequency components below some cutoff frequency, f_c. An example highpass application is cutting out the audible 60-Hz AC power "hum", which can be picked up as noise accompanying almost any signal in the United States.

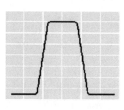

The designer of a cell phone or any other sort of wireless transmitter would typically place an analog *bandpass* filter in its output RF stage, to ensure that only output signals within its narrow, government-authorized range of the frequency spectrum are transmitted.

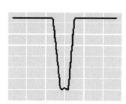

Engineers can use *bandstop* filters, which pass both low and high frequencies, to block a predefined range of frequencies in the middle.

3.5   COMPARING ANALOG WITH DIGITAL FILTERS

Most digital signals originate in analog electronics. If the signal needs to be filtered, is it better to use an analog filter before digitization, or a digital filter after? We will answer this question by letting two of the best contenders deliver their blows.

The flatness achievable with analog filters is limited by the accuracy of their resistors and capacitors. Analog filters of high complexity will have a residue ripple of, perhaps, 1%. On the other hand, the flatness of digital filters is primarily limited by round-off error, making them *hundreds* of times flatter than their analog counterparts. Score one point for the digital filter.

Next, the digital filter is also the victor in both *roll-off* and *stopband attenuation*. Even if the analog performance is improved by adding additional stages, it still cannot compare with the digital filter. For instance, imagine that you need to improve these two parameters by a factor of

100. This can be done with simple modifications to the windowed-sinc, but is virtually impossible for the analog circuit. Score two more for the digital filter.

The digital filter's step response is symmetrical between the lower and upper portions of the step, i.e., it has a linear phase. The analog filter's step response is *not* symmetrical, i.e., it has a non-linear phase. One more point for the digital filter.

In spite of this beating, there are still many applications where analog filters should, or must, be used. This is not related to the actual performance of the filter (i.e., what goes in and what comes out), but to the general advantages that analog circuits have over digital techniques. The first advantage is *speed*: digital is slow; analog is fast. For example, a personal computer can only filter data at about 10,000 samples per second, using FFT convolution. Even simple op amps can operate at 100 kHz to 1 MHz, 10 to 100 times as fast as the digital system!

The second inherent advantage of analog over digital is *dynamic range*. This comes in two flavors. *Amplitude dynamic range* is the ratio between the largest signal that can be passed through a system and the inherent noise of the system. Just as before, a simple op amp devastates the digital system in this domain.

The other flavor is *frequency dynamic range*. For example, it is easy to design an op amp circuit to simultaneously handle frequencies between 0.01 and 100 kHz. When this is tried with a digital system, the computer becomes swamped with data. For instance, sampling at 200 kHz, it takes 20 million points to capture one complete cycle at 0.01 Hz.

3.6   OPERATIONAL AMPLIFIER

As its name indicates, operational amplifier can be used to perform mathematical operations on voltage signals such as inversion, addition, subtraction, integration, differentiation, and multiplication by a constant.

One of the cheapest and most popular OP-AMP is the 741. Here, the 741-op-amp will be used to add signals together.

Let us take a look first at the pinout configuration of the op-amp:

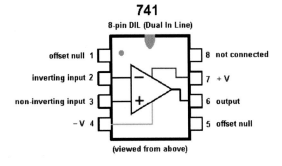
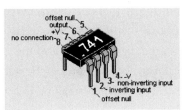

The usual circuit symbol for an op-amp is:

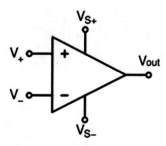

V_+: non-inverting input, V_-: inverting input, V_{out}: output, V_{S+}: positive power supply (sometimes also V_{CC+}), V_{S-}: negative power supply (sometimes also V_{CC-})
The output voltage (V_{out}) cannot be higher than V_s.

3.7 EXPERIMENT

Now that you have an idea about analog filters, you are ready to begin the experiment.

3.7.1 Part A. Low Pass Filter

Launch NI ELVIS. On the workstation protoboard, build a simple RC circuit, as shown in the following schematic diagram, using $C = 0.1$ uF and $R = 3.3$ kΩ.

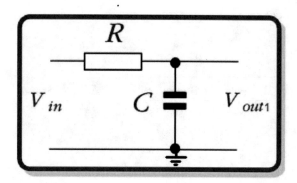

Using the function generator panel, set V_{in} to have the following parameters:

- Waveform: Sine wave
- Peak amplitude: 2 V p-p
- DC Offset: 0.0 V

Set the input frequency to 500 Hz and verify that the output amplitude is 0.707*input amplitude.

1. Compare 500 Hz with the theoretical value of the cutoff frequency for the resistor and capacitor used. Hint: $f_L = 1/(2\pi RC)$.
What happens when you vary the frequency?

A fast way of studying the frequency response curve of the Lowpass Filter is to use the ELVIS Bode plot feature. The Bode plot is basically a plot of Gain (dB) and Phase (degrees) as a function of log frequency. Complete the following steps to measure the Bode plot of the RC circuit:
  –From the NI ELVIS Instrument Launcher, select *Bode Analyzer*.
  –Connect the signals, input (V_1), and output (V_{out}), to the analog input pins as follows:
  V_{in} (FUNC_OUT) to ACH1+ and Ground to ACH1–
  V_{out} to ACH0+ and Ground to ACH0–
  –From the Bode Analyzer, set the scan parameters as follows:

  •   Start: 10 (Hz)
  •   Stop: 10,000 (Hz)
  •   Steps: 10 (per decade)

  Switch the Function Generator to AUTO mode.
  –Click *Run*

2. Describe the Bode plot.

3. Use the cursor function to find the low frequency cutoff point, which is the frequency at which the gain has fallen by –3 dB. Compare it with $f_L = 1/(2\pi RC)$

3.7.2 Part B. High Pass Filter

On the workstation protoboard, connect a 100–Ω resistor and the 0.1-μF capacitor to have the following high pass filter circuit:

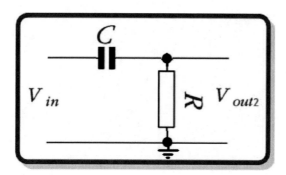

Set the input frequency to 16 kHz and verify that the output amplitude is 0.707*input amplitude.

4. Compare 16 kHz with the theoretical value of the cut-off frequency for the resistor and capacitor used. Hint: $f_U = 1/(2\pi RC)$.
What happens when you vary the frequency?

5. Run Bode plot using the same scan parameters as before and describe the resulting bode plot.

6. Use the cursor function to find the high frequency cutoff point, that is, the frequency at which the amplitude has fallen by −3 dB. Compare it with $f_U = 1/(2\pi R_U C_U)$.

3.7.3 Part C. Summing Signals

If a low-pass filter is combined along with a high-pass filter, the resulting filter is band-pass or band-stop filter.

Construct the circuit below with your choice of R. The circuits adds V_1 and V_2 together, $V_{out} = -(V_1 + V_2)$. Use $V_{cc} = +15$ V

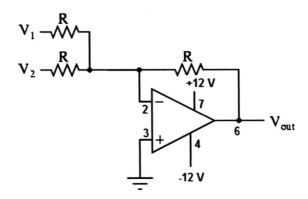

Test your circuit by connecting to V_1 and V_2 +5 V and +5 V, respectively, and measure the output voltage. Does it correspond to $-(V_1 + V_2)$?

Replace V_1 and V_2 by the output of the low-pass and high-pass filters as shown below.

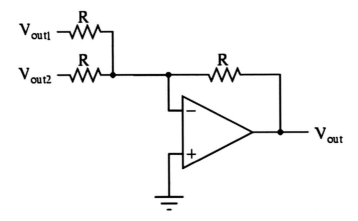

7. Run the Bode plot using the same scan parameters as before and describe the resulting bode plot. What is the resulting filter?

8. Use the cursor function to find the low and high frequency cutoff points at which the amplitude has fallen by −3 dB. Compare it with $f_L = 1/(2\pi R_l C_l)$ and $f_U = 1/(2\pi R_U C_U)$.

9. Replace the input of the low-pass and high-pass filters by the audio corrupted signal. Is the output signal still corrupted? Explain why.

•   •   •   •

CHAPTER 4

Music Equalizer Using Op-Amps: Volume and Treble Control

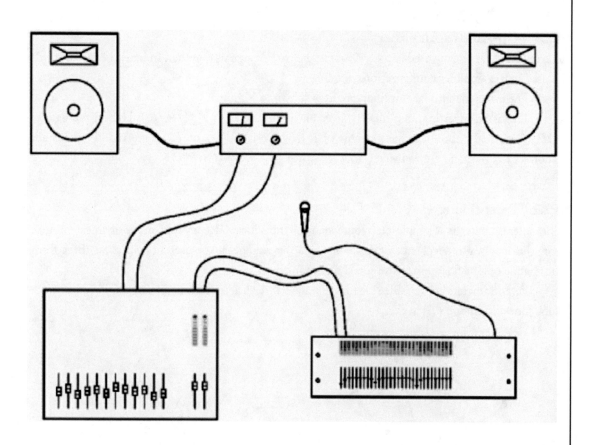

4.1 OVERVIEW

In this chapter, you will use analog components such as op-amps to control the treble frequencies of music.

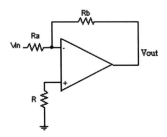

FIGURE 4.1: Inverting amplifier.

4.2   BACKGROUND

4.2.1   Op-Amp as Amplifier

The op-amp was already used in a previous chapter to add signals together. In here, the op-amp will be used in the amplifier configuration.

The basic operational amplifier circuit is shown in Figure 4.1.

This is an inverting amplifier. It gives an output of $V_{out} = V_{in}*(-R_b/R_a)$. Hence, the gain is R_b/R_a. R is usually chosen to have a value of $R_a||R_b$. Resistors R_a and R_b connected in parallel are denoted by $R_a||R_b$ and are equivalent to a resistor $R = (R_a R_b)/(R_a + R_b)$.

4.2.2   Potentiometers

The potentiometer is a simple electromechanical transducer. It converts rotary or linear motion from the operator into a change of resistance, and this change can be used to control anything from the volume of a hi-fi system to the speed of a motor.

Potentiometers are variable resistors that have three leads. To connect them, follow the directions shown in Figure 4.2.

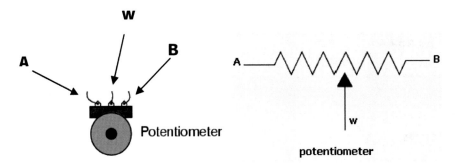

FIGURE 4.2: Potentiometer connections.

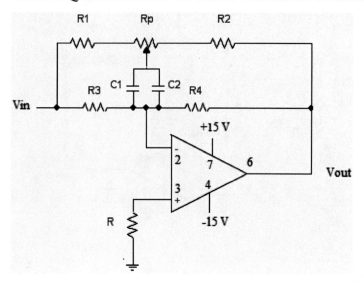

FIGURE 4.3: Treble control.

4.2.3 Treble Control

The classic form of tone control in hi-fi is the Baxandall arrangement [1] shown in Figure 4.3.

This arrangement is called a Baxandall treble control, named after its inventor [1]. We can understand how it works by noticing that it is actually a development of the inverting amplifier arrangement shown previously. However, the normal pair of input and feedback resistors has been replaced by quite complicated arrangements of resistance and capacitance. The circuit is laid out in a symmetric manner.

At low frequencies, the pair of capacitors act as an open circuit, which means the treble control circuit is equivalent to the one shown in Figure 4.4.

However, at higher frequencies, the capacitors act as a short circuit, which means the treble control circuit is equivalent to the one shown in Figure 4.5.

Hence, the R_p pot acts as a Treble control and allows us to boost or cut the relative gain for high frequency signals.

With the pot turned fully to one end, the maximum gain is obtained

$$G_{max} = -\frac{R_4||(R_p + R_2)}{R_1||R_3}$$

With the pot turned fully to the other end, the minimum gain is obtained

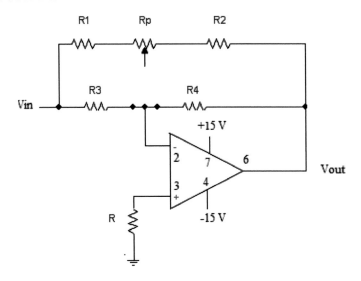

FIGURE 4.4: Equivalent treble control circuit at low frequencies.

$$G_{min} = -\frac{R_2||R_4}{R_1||(R_p + R_3)}$$

Hence one can boost and cut the high frequencies by turning the potentiometer in both directions.

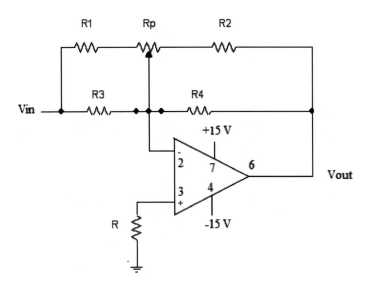

FIGURE 4.5: Equivalent treble control circuit at high frequencies.

Components:

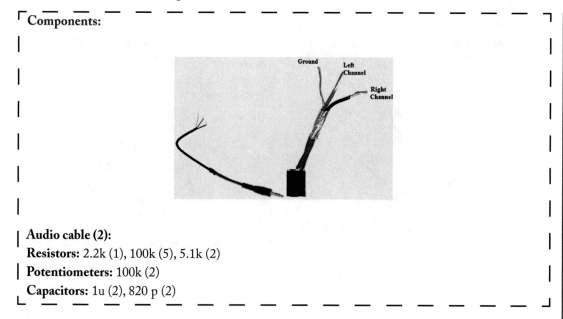

Audio cable (2):
Resistors: 2.2k (1), 100k (5), 5.1k (2)
Potentiometers: 100k (2)
Capacitors: 1u (2), 820 p (2)

4.3   EXPERIMENT

Now that you are familiar with op-amps, you are ready to begin the experiment.

4.3.1   Volume Control

When designing a music equalizer, the user has also to be able to control the volume. The following circuit serves that purpose.

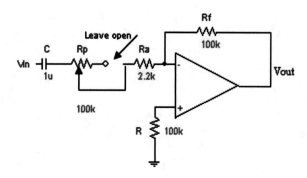

Construct the circuit while making the right connections on the ELVIS workstation.

4.3.2 Help Desk

Potentiometers are variable resistors that have three leads. To connect them, follow these directions:

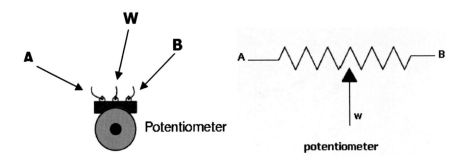

Using a sine wave of 1 kHz and 100 mVp as input, measure the peak-to-peak amplitudes of *both* the input and output signals at *both* limit positions of the potentiometer corresponding to the minimal and maximal gain.

Compute the gain with $R_f/(R_a + R_p)$ where:
  1. $R_p = 0$
  2. $R_p = 100k$

Which resistance corresponds to max gain, and which corresponds to min gain? Do they match the experimental values obtained?

4.3.3   Music Equalizer: Treble Control

The next step is to construct the treble control. While keeping the previous circuit constructed, connect the treble control circuit as follows.

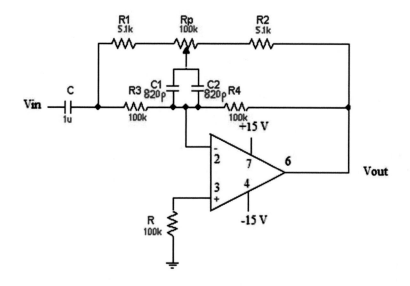

Construct the circuit above, using ±12 V supplies with $R_S = 1$ kΩ, $R_1 = R_2 = 5.1$ kΩ, $R_3 = R_4 = 100$ kΩ, $R_p = 100$ kΩ, $R = 100$ kΩ, $C1 = C2 = 820$ pF, and $C = 1$ μF. Add a load resistor RL = 1 kW at the output.

Set the frequency to 20 kHz, and turn the potentiometer wheel to one extreme position then to the other. What do you notice about the output? Explain.

> Set the frequency to 20 Hz, and turn the potentiometer wheel to one extreme position then to the other. What do you notice about the output? Explain.

Hint: Turning a treble control knob should not alter the bass tones!

4.3.4 Combining Volume and Treble Control

Connect a CD player instead of the function generator as follows:

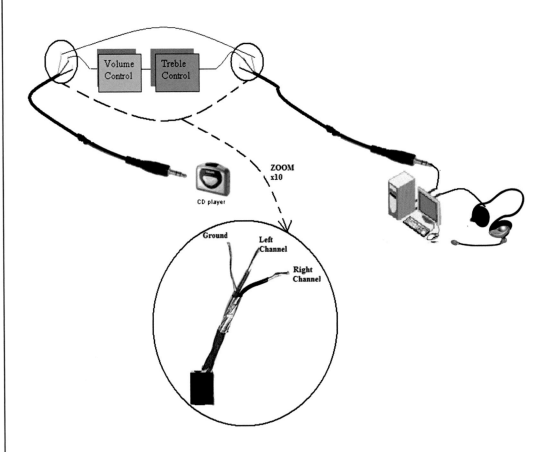

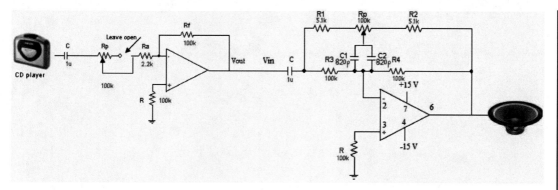

Vary the volume control and the treble control knobs (potentiometers) and notice the change in the music. Comment.

REFERENCES:

[1]   http://sound.westhost.com/dwopa2.htm.

C H A P T E R 5

Music Composer Using 555 Timers

5.1   OVERVIEW

In this chapter, you will be using analog components such as 555 timers to generate the musical tones of a piano.

5.2   BACKGROUND

5.2.1   555 Timer

The 555 timer is one of the more versatile integrated circuits ever produced. It can be used to build lots of different circuits. It contains 23 transistors, 2 diodes, and 16 resistors inside a very small package with eight legs type DIP (dual-in-line package), as shown below:

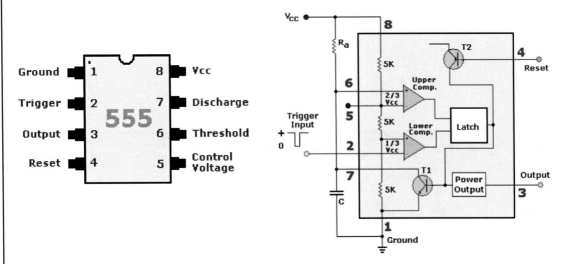

The figure above on the right is what is inside the 555 timer chip. The 555 timer contains electronic micro-components made up with semiconductor materials. That is why it is called an integrated circuit.

5.2.2   Oscillator Circuit

Astable circuits are the configurations in which 555 timers produce pulses, like a digital clock. The circuit most people use to make a 555 astable circuit looks like this:

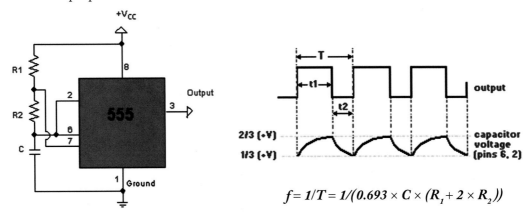

$$f = 1/T = 1/(0.693 \times C \times (R_1 + 2 \times R_2))$$

To have $t_1 = t_2$, use $R_1 = 0$.

Timer circuit. The 555 can be used also as a timer in the monostable mode. The output of the timer is normally low; if the trigger button is pressed or goes low, then the output goes high for $t =$ 1.1RC seconds.

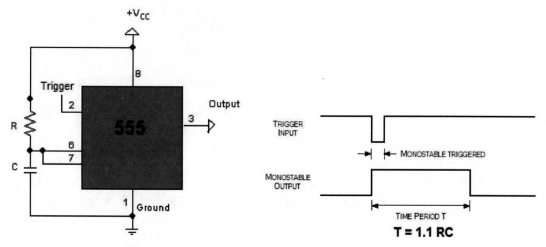

5.3   EXPERIMENT

Now that you got familiar with the operation of 555 timers, you are ready to begin the experiment.

5.3.1   Music Composer

The following circuit simulates different musical tones:

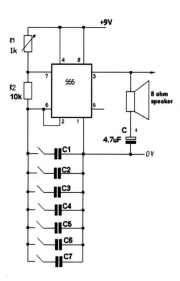

Construct the circuit shown above, using the capacitor values shown in the following table:

	C(µF)
C1	.100
C2	.068
C3	.047
C4	.033
C5	.022
C6	.015
C7	.010

Note: The capacitor with value 4.7 uF is electrolytic. This means that it has a polarity and has to be connected taking care of where the positive and negative terminals go; otherwise, it will pop and burn!

Calculate the frequency of the generated tone corresponding to each capacitor using:
$$f = 1/T = 1/(0.693 \times C \times (R_1 + 2 \times R_2))$$

Try playing different tones by switching different capacitors on and off.

Author Biography

Lina J. Karam received her BS degree in engineering from the American University of Beirut in 1989, and earned her MS and PhD degrees in Electrical Engineering from the Georgia Institute of Technology in 1992 and 1995, respectively. She is currently an associate professor in the Department of Electrical Engineering at the Arizona State University. She is currently the director of the Image, Video, and Usability, the Multi-Dimensional DSP, and the Real-Time Embedded Signal Processing laboratories in the Department of Electrical Engineering at ASU. Her research interests are in the areas of image and video processing, compression, and transmission; human visual perception; multidimensional signal processing; error-resilient source coding; digital filter design; and biomedical imaging. From 1991 to 1995, she was a research assistant in the Graphics, Visualization, and Usability Center and then in the Department of Electrical Engineering at Georgia Tech. She has worked at Schlumberger Well Services (Austin, TX), and in the Signal Processing Department of AT&T Bell Labs (Murray Hill, NJ) in 1992 and 1994, respectively. Prof. Karam is the recipient of a US National Science Foundation CAREER Award. She served as the Chair of the IEEE Communications and Signal Processing Chapters in Phoenix in 1997 and 1998. She was a member of the organizing committees of the 1999 IEEE International Conference on Acoustics, Speech, and Signal Processing (ICASSP99), the 2000 IEEE International Conference on Image Processing (ICIP00), the 2008 IEEE International Conference on Acoustics, Speech, and Signal Processing (ICASSP08), and Asilomar 2008. She is the Technical Program Chair of the 2009 IEEE International Conference on Image Processing (ICIP09). She is an associate editor of the *IEEE Transactions on Image Processing* and is serving on the editorial board of the *Foundations and Trends in Signal Processing Journal*. She also served as an associate editor of the *IEEE Signal Processing Letters* from 2004 to 2006, and as a member of the IEEE Signal Processing Society's Conference Board. She is an elected member of the IEEE Circuits and Systems Society's DSP Technical Committee, and of the IEEE Signal Processing Society's Image and Multidimensional Signal Processing Technical Committee. She is a senior member of the IEEE and a member of the Signal Processing and Circuits and Systems societies of the IEEE.

Naji Mounsef received his BS degree (magna cum laude) in computer and communication engineering from Notre Dame University (NDU), Lebanon, in 2004, and his MS degree in computer and communication engineering from the American University of Beirut (AUB) in 2005. While at NDU, he served as a teaching assistant in circuits, electronics, logic design, and digital signal processing laboratories. From 2004 to 2005, he was also a teaching assistant at the AUB in digital signal processing and digital image processing laboratories, for which he wrote a laboratory manual. During the same period, he worked as a research assistant at AUB and worked on problems related to software radio and efficient turbo decoding. After teaching for a semester at NDU and AUB, he joined the PhD program in Electrical Engineering (2006) at the Arizona State University (ASU), where he first worked as a research assistant and then as a teaching assistant. In summer 2008, he interned at the Translational Genomics Institute. He is part of the Image, Video, and Usability (IVU) laboratory in the Electrical Engineering Department of ASU, and his research areas include image processing and genomic signal processing.

Printed in the United States
by Baker & Taylor Publisher Services